Power Particle

Power Particle

Raton – The Key to Universal Energy and Beyond

Dr. Bashamber N. Chabra

POWER PARTICLE

Copyright 2024 © Dr. Bashamber N. Chabra

All information, techniques, ideas and concepts contained within this publication are of the nature of general comment only and are not in any way recommended as individual advice. The intent is to offer a variety of information to provide a wider range of choices now and in the future, recognizing that we all have widely diverse circumstances and viewpoints. Should any reader choose to make use of the information contained herein, this is their decision, and the contributors (and their companies), authors and publishers do not assume any responsibilities whatsoever under any condition or circumstances. It is recommended that the reader obtain their own independent advice.

First Edition 1998
Second Edition 2024

ISBN: 9798335924108

All rights reserved in all media. No part of this book may be used, copied, reproduced, presented, stored, communicated or transmitted in any form by any means without prior written permission, except in the case of brief quotations embodied in critical articles and reviews.

The moral right of Dr. Bashamber N. Chabra as the author of this work has been asserted by him in accordance with the Copyrights, Designs and Patents Act of 1988.

Published by Happy Self Publishing
www.happyselfpublishing.com
writetous@happyselfpublishing.com

This book is dedicated to the love of my life, my wife Daya; in loving memory of my parents, Hans Raj and Inder Kaur; and last but not least, to our benevolent and ever-so-giving, Mother Earth.

Acknowledgments

I would like to thank my wife, Daya, for her moral support and my daughters, Ratna, Renu, Sangeeta, and Poonam, for editing the book and for getting it on its feet.

I would also like to share my gratitude to Rama Samudrala for his crucial advice, without which this book would have been incomplete. As a result, I was able to refine the book to its fullest potential.

Table of Contents

Note to the Reader 1
Introduction 3

Part One: Exploring the Raton 7

1. The Raton and Its Cycles..9
2. Evidence of the Raton in Past Scientific Theories ..31

Part Two: Understanding the Universe 35

3. The Earth – It's Alive 37
4. The Genesis of the Solar System 49
5. The Final Events of the Solar System....................59
6. The Asteroid Belt...67

Bibliography 73
About the Author..75

Note to the Reader

Dear Reader,

This book was originally published in 1998. With a goal to reach a wider audience in the digital age, I am re-releasing it. I invite you to explore and expand upon the concepts as you see them relevant to current times. Wishing you a blessed life in this magnificent universe that we live in.

Introduction

The history of science demonstrates that one era of scientists starts a discovery, breaking the ground and making it fertile for future scientists to pick up where their predecessors left off. Every so often, a culmination of knowledge manifests in a historical breakthrough that utilizes the information gathered over past centuries and serves as a landmark. The first printing press, the first steam engine, the first automobile, the first man on the moon are all examples of these landmarks. These landmarks were reached by small steps made by countless people, many of whom are unknown or forgotten. Some of the ideas that led to great discoveries were discounted as valid after many years, decades, and even centuries. So it has been and will continue to be.

This book will attempt to carry the torch of science one leg further. The theories introduced here rely on the foundation of past findings and aspire to strengthen the basis that future minds will expand on.

For centuries, scientists have attempted to understand and describe the ultimate elementary particle that comprises the basic building block of all matter existing in the universe. Once upon a time, scientists concluded that the atom was the ultimate particle. Later, scientists realized that there were yet smaller particles, which together formed the atom. Albert Einstein provided a superb explanation of the ultimate composition of all matter. He theorized that all matter is merely a form of energy and explained that matter and energy are constantly interchanging. Einstein's theory, however, did not explain the mechanics of the interchange. That's where this book takes up the torch.

Here, I substantiate Einstein's theory while expanding it to include an explanation of how and why matter ultimately becomes energy and vice versa. I also attempt to define the ultimate elementary particle, which is the basic building block of all matter in the universe. I have called this particle "raton." This particle is present in various phases and is the very fundamental component that makes up energy and matter.

In the next few chapters, I will explain the basis of my theory, how I derived it, and what it is comprised of. I will also explain the impact that the raton theory can make in our present understanding of the universe and its formation.

I will demonstrate that everything in nature is part of a forever-circulating ratonic cycle-which moves from a dormant phase to activated forms, such as electromagnetic waves, fire, heart, and matter. Once the ratonic cycle is grasped, the enigma of how nature works will begin to unfold.

Part One

Exploring the Raton

1

The Raton and Its Cycles

For centuries, learned people have sought the answers to nature's guarded secrets, with one of the most elusive questions being, "What is the ultimate particle of the universe?" Scientists continue the search for a substance that is indivisible and is the very makeup of the entire universe. The raton particle is that substance.

The raton is nature's smallest particle. It has no further subdivision. All matter in the universe is made up of it. All matter in the universe eventually reverts back to it. The raton is thermostable, electrically neutral, chemically inert, and free of gravitational influences. The raton is the basis of all matter, yet, singularly, its own mass is diminutive to such an extreme that it may be perceived as massless. Unlike the atmosphere, it exerts no pressure. It permeates everything, and a vacuum cannot be created. It is ubiquitous.

Ratons work in cycles, traveling through various forms of stages. Within the cycle, ratons are converted into matter, which eventually reverts back into ratons. To illustrate the cycles, I will refer to a log of wood. The breakdown of a log of wood is easy to comprehend because it happens quickly and at a temperature that is feasible in our atmosphere. When one observes a log on fire, one sees the flickering flames, feels warmth, sees light, and sees smoke rising upwards. After some time, the flames subside as the log is consumed and a small amount of ash is left behind. The room will cool and the light and smoke will have disappeared.

If it were possible to collect all the smoke, ashes, gases, and vapor that resulted from this process then compress and weigh them, the weight of the by-products would be more than the original weight of the log of wood. This is because of the addition of Oxygen, which combined with the carbon from the wood to form carbon dioxide. If the weight of this Oxygen was to be subtracted from the total weight of the by-products, then the new weight should be less than the original log of wood. Hence, it follows that there is a loss of mass (*see Discusion-1*). This small loss of mass converts into energy, which is the basis of my following discussion.

DISCUSSION-1

In 1772, a French chemist by the name of Antoine Laurent Lavoisier discovered the chemical principles of combustion. He performed an experiment where he took some accurately-weighed sulfur and burned it under controlled conditions. His goal was to collect the resulting smoke and vapor in order to obtain the weight. He found that the resulting by-products actually weighed more than the original sample of the sulfur. It thus followed that the extra weight was a result of the addition of Oxygen during combustion.

Similarly, the "Burning Candle Experiment" demonstrates that when a candle is burnt under controlled conditions, there is a conversation of matter in different forms. Like Lavoisier's results, the by-producers of the burnt candle weighed more than the candle itself. Once again, it is obvious that the incremental increase of weight is attributed to the ingress of the Oxygen.

It is not possible to remove Oxygen from the by-products in the above experiments. Hypothetically, if the

> *weight of the Oxygen were to be subtracted from the sum of the by-produces, the remaining weight would be less than the weight of the original source, i.e., the fission, as performed by nuclear physicists, supports this hypothesis.*
>
> *In the fission of radioactive material, like the atomic blast, when a neutron passes through a critical mass of fissionable uranium, there is a profound release of energy in a matter of a split second. The leftover products from the uranium will weigh less than the original fissionable mass. Albert Einstein explained this conversion of mass to energy within the formula: energy = (mass) (speed of light)2 or $E = MC2$.*

We know that mass is never destroyed and that it only changes its form. Albert Einstein's theory of relativity tells us that mass and energy are interrelated and transferable. What are the mechanics of the transfer that takes place? We are witnessing the different phases of the ratonic cycle. In order to understand the concept, I will take you through each of the five phases (Phase 4 to Phase 0) of the ratonic cycle.

PHASE 4 – MATTER

Phase 4 represents ratons in their most solid form, matter. Matter exists throughout the universe but little is known about its formation. Matter may be divided into two broad categories: (1) matter that is dependent on photosynthesis, which includes all human, animal, and plant life; and (2) matter that is not covered by the photosynthesis process, which will be referred to as geo-matter.

Matter is made up of various combinations of elements. Spectroscopic research of stars reveals that stars also have the same elements. How these elements came into existence will be discussed in later sections.

For the purposes of our illustration, matter is represented by the log of wood before it was set on fire.

PHASE 3 – FIRON

Phase 3 is represented by fire. Although fire is an essential of daily life, we know very little about it. We know it is comprised of flames and causes heat, light, and smoke, but what is the flame?

Flames are made up of minute particles that I will call "firon." These particles are extremely diminutive – there would be trillions of them in a square micrometer film of the flames. The firon is crimson red. It is lighter than the atmosphere and, therefore, rises upwards. At room temperature it is unstable and breaks down into its two component parts, heat, and Oxygen. It is stable at extremely high temperatures that can be achieved in the core of the Earth and the corona of the sun, where the temperatures are approximately five thousand degrees Kelvin and one hundred thousand degrees Kelvin, respectively.

Firons tend to flow from high temperatures to low temperatures. For example, on the corona of the sun, there are flares of flames that shoot thousands of miles outwards. This outpouring of flames is governed by the temperature differential surrounding the sun's corona.

The firon particle is composed of a bond between oxygen and heat particles, or "heatrons," which will be elucidated later in this chapter. This is the reason that fire cannot be started without Oxygen. [Flame (firon) = heat particles (heatron) + Oxygen]. In this phase, the loss of volume of the log of wood is being converted into firon particles.

In addition to heat particles and Oxygen, to start a fire, one needs an **optimal temperature**. The **optimal tem-**

perature or **OT** of a substance is the temperature that will facilitate a chemical reaction to take place between the substance and the Oxygen.

As a result of this reaction, firon particles can form. Let's take an example of a sheet of paper on the ground where the sun can shine on it. The paper will not catch fire if it does not reach its OT. If you place a magnifying glass over it so that sun rays are concentrated on a small area of the paper, the paper will ignite. This indicates that the concentrated sun rays increase the temperature on the paper until it reaches its OT. The OT serves as a catalytic agent to start a chemical reaction between Oxygen and the quantum particles of the paper to convert them to the firon particle. The optimal temperature varies depending on the substance. On a very hot day, grass may ignite spontaneously. The tree branch may not ignite because the grass and the branch have different OT. As we know, petroleum ignites very quickly. That is because of its extremely low OT.

All matter is combustible if it reaches its OT. It was previously believed that the 103 elements on Earth were noncombustible. The ratonic hypothesis states that they are indeed combustible, as is all matter, but their OT is extremely high and not possible on this planet. At temperatures of millions or billions of degrees Kelvin, as seen in supergiant stars, these elements would undergo the chemical reaction with Oxygen to form firon molecules. At those temperatures, the elements are susceptible to the same cycle as the log of wood in our illustration. Even the planet Earth would burn like a log of wood at its OT.

PHASE 2 – PHOTON

Phase 2 is made up of electromagnetic, gamma, roentgen, ultraviolet, micro, and radio waves. The members of this phase are involved in energy transportation.

Light is the most prominent member of this phase. It is a fast energy-transporting medium that travels from the sun to the Earth at a distance of 93 million miles in just 8½ minutes. The structure of a ray is still unclear. The famous English scientist Isaac Newton believed that light consists of minute particles, whereas Christian Huygens, the Dutch astronomer and physicist, described light as a wave phenomenon. In 1928, Niels Bohr, a Danish physicist, described the complementary principle based on an experimental arrangement, which suggested that light is sometimes particle-like and sometimes wavelike; that is, it has a wave-particle duality.

The Ratonic Theory suggests that light is a fast-moving jet of particles that rises from an atom as a spiral. Light travels at such a tremendous speed that the spirals spinning from the atoms are interpreted as waves. This spiral is the key to understanding the roles of various members of this phase. If light travels at 186,000 miles/second, the absolute velocity of the photons would be a thousandfold more than 186,000 miles (*see Discussion-2*).

During the burning of the wood in our illustration, a portion of the log's volume is dispersed as light, which is composed of photons.

DISCUSSION-2

Light has a wavelength of 4000-7000 angstroms. The wavelength is the distance from the crest of one wave to the crest of a second wave. An Angstrom unit is one hundred-millionth of a centimeter or two hundred-fifty-millionth of an inch. Thus, there are 35,700-62,500 spirals in one linear inch of a light ray. If one inch of the spiral could be stretched so that the photons would be in a straight line, imagine the length of a ray of light. I believe that photons of the spiral would line up to be many hundreds or thousands of times greater than the original length.

The gamma ray wavelength is 40 to 70 angstroms. In one linear inch of gamma ray, there will be around 1,000 times more spirals than that of light and likewise the absolute velocity of gamma ray photons would be around 1,000 times that of light photons. These high velocity photons can pass through the human body without any resistance, causing considerable tissue damage. If one were to thrown a hand-held bullet at a person, no harm would be done, but if the same bullet is fired through a pistol, it can cause considerable tissue damage. These very high-velocity traversing photons can cause serious levels of morbidity, such as seen in leukemia and other malignancies. In August 1945, after the blast of the atom bomb in Hiroshima and Nagasaki, Japan, many people near the hypocenter who survived the blast developed leukemia. The high velocity photons from the blast that penetrated the human bodies destroyed the ultra-microscopic structure of the bone marrow.

The white blood corpuscles have the function of protecting against infections. The normal count is from 4,000 to 11,000 wbc per cubic millimeter of blood. When a person is infected with any infections, such as tonsillitis or appendicitis, the body mechanism that controls the productions and maturations of these cells augments their production and the mature wbc in blood are increased by 25 percent or so. When the high velocity photons of the radioactive material pass through the human body in substantial doses, they destroy the control mechanisms of the body. This results in a marked increase in the production of immature cells in the bloodstream. This cell count can go from 50,000 to 100,000 cells. These immature cells cannot protect the host, who consequently dies due to the lack of immunity.

PHASE 1 – HEATRON

As our log of wood is consumed by fire, the firon particles continuously break down into their component parts; Oxygen and heat. The heat is now devoid of Oxygen and made up of particles that I will call "heatron" or "heatron complexes."

Let's look for a moment at mammals. In order to maintain good health, mammals must regulate their body heat within a certain range continually. This maintenance of body heat is dependent upon the concentration of heatron complexes that are distributed evenly in each and every cell among the billions of cells in the body.

Heatrons are responsible for burns. When the firon molecules come into contact with skin, it breaks into heatrons and Oxygen. The heatrons are absorbed into the skin and cause the burn. It is interesting to observe people who walk on fire and remain unscathed. Possibly, their body secretes a hormone that negates the breakdown of the flame into heatrons and Oxygen, thereby avoiding the burn injury. The firon particles, per se, are inert.

Heat is obvious evidence of the presence of heatron complexes. Temperature is the measurement of their concentration. These heatron complexes disintegrate to ratons, the most basic particles.

These heatrons flow from higher concentrations to lower concentrations, or we can say from higher temperature to lower temperature. For example, when one inserts a mercury thermometer in the mouth of a febrile person, the mercury in the glass tube expands to indicate the person's temperature. In the past, this phenomenon was interpreted as the expansion of the metal when heat is applied. Now, we go one step further. The expansion of the mercury is due to the absolute physical entry of the heatron particles from the person's mouth to the mercury.

Like water that flows from higher to lower levels until it equalizes, heatrons flow from higher to lower concentrations. However, water equalizes because it is governed by gravity, whereas, heatron complexes are not susceptible to gravity, but ruled only by their nature to flow from high concentration to low concentration.

The heatron particles' collective force can be illustrated by a simple experiment. We perceive that metal is a very sold substance, but from the heatronic point of view, it is actually very porous. Let's observe an iron rod that is being heated at one end. First, note that the rod becomes hot throughout its length. This is known as the conduction of heat by the metal. Heatron particles rush into the pores of the metal as water passes through a sponge. In a short time, the heatron fills the pores of the metal and causes the temperature of the entire rod to increase.

Second, the metal will expand and lengthen. The heatron complexes fill the pores of the metal and a small portion of heatrons exit through the pores. That is why one can feel the heat of the rod from a distance. The exodus of the heatrons takes place at a slower rate than its entry. Therefore, the heatrons get crowded and they perforce overpower the resistance of the substance of the iron rod and start pushing its minute structure apart.

That causes the entire volume of the rod to increase, thus the expansion that we observe. Eventually, the rod will become red-hot. The color is due to the firon particle. As the heatron expands the pores of the metal, the pores become wide enough for the Oxygen to enter and join together with the heatron complexes. At this point, a reversal phenomenon occurs in which the heatrons attract Oxygen thereby forming firons (Heatron + Oxygen = Firon).

If through the process of heating, the metal were to reach its OT, it would ignite. However, that wouldn't happen because iron's OT is extremely high. The metal will continue in its present state until the source of heat is removed. As we continue to apply heat, the rod will begin to melt. This occurs because the metal pores expand to such an extent that the wall of the pores cannot hold the structure intact. At this stage, the metal's volume is increased considerably. When the heat is removed and the iron bar starts to cool, the crowded heatrons continue to emanate from the iron, but a fresh supply is no longer being provided by an external heat source. The heatrons in the iron start reducing in numbers and the elasticity of the metals starts dominating the heatrons, and ultimately, the rod regains its original size and form. However, if the rod is deformed during heating, it will maintain that form.

This phenomenon can also be seen in everyday things like telephones and electric wires that stretch from pole to pole on our streets. The wires are generally lax in the summertime because of their increased length and taut in the winter months. Also, railway tracks are placed a little apart at their joints so that they don't curve during the heat of summer.

Another interesting observation of the role of heatrons in our everyday life is apparent in our cooking. Let's take for example, cooking rice. When 2 cups of rice are boiled with 3 cups of water, is should follow that the volume of cooked rice would be 5 cups. Instead, the process will result in 6 to 7 cups of cooked rice. This is the result of the actual physical entry of heatrons into each grain of rice. The heatrons flush the pores of the grains and expand them. Unlike the iron rod, rice is not elastic and, therefore, maintains the expanded volume even when it is cooled.

To further illustrate this point, let's consider the formation of water. Water can be formed by applying heat to two adjoining cylinders, one containing hydrogen and the other holding Oxygen. It has been believed that water is a compound of hydrogen and Oxygen. But the Ratonic Theory states that it is actually a compound of hydrogen, Oxygen, and heatrons. Heatrons are released along with hydrogen and Oxygen when water is separated by electrolysis. The heatrons that escape are never accounted for in chemistry reactions, but they are always present.

Let's refer to our log of wood. It has been transformed from matter (Phase 4), to firon (Phase 3), to photons (Phase 2), and to heatrons (Phase 1). The Ratonic Theory suggests that the loss of mass from the wood was due to the escape of heatron complexes. If it were possible to collect the ashes, smoke, heatrons, and photons, the mass would equal the mass of the log before it was burned.

PHASE 0 – RATON

The final destination of our log of wood, like all matter, is the raton. The transformation of heatron complexes to ratons is the most basic in nature. Heatrons may be viewed as an activated raton, whereas ratons can be viewed as the resting phase of heatrons. As the concentration of heatrons decreases, causing a drop in temperature, it reaches a point of low concentration where the structure of heatron is dismantled. The reason for this breakdown is for future generations to work out with technology that is not available in our time. For now, the best way to explain the heatron complex is as a coded single-strand ratonic structure in motion. When the speed is reduced to certain limits, the heatron disintegrates and ends up as ratons. The ubiquitous ratons are the source and the destination of everything in the universe. The planets and stars are orbiting within an "ocean" of ratons. The ratonic ocean is the medium for the propagation of the electromagnetic waves.

Contraption of Energy

From our log of wood illustration, we saw how matter is broken down to its most fundamental unit, the raton. Let us explore the process in which ratons evolve into energy. This process shall be called **contraption of energy**. The contraption of energy takes place with the help of the atom. An atom is a hollow structure with a central nucleus. In the hollow space in between the nucleus and the wall of the atom, there are still smaller particles, the electrons. The nucleus is made up of protons and neutrons, and other particles such as the mesons, neutrinos, hyperons, and antiparticles. The atom is so diminutive that a million atoms piled one atop the other would make a column that might equal the thickness of this page. The electron mass would be 0.000000000000000000000000009 grams or 9^{-27}.

Imagine the atom as a solar system, the nucleus as the sun, and the electrons as planets that orbit around the sun/nucleus. The electrons move at a speed of about a billion orbits around the nucleus in a millionth of a second. According to Danish physicist Niels Bohr, when the electron is excited and changes its position from high to intermediate levels, energy is given off.

The Ratonic Theory suggests that the atom is a miniature factory unit, and its hollow space is filled with ratonic particles or "raw material." Just as a spindle changes raw material, like cotton, to yarn, the electron's fast orbit around the nucleus changes the latent "raw material," the raton particles, into a ray of the radiant energy (Phase 2). The ray of radiant energy leaves the atom in linear form. Since the electron is orbiting in tremendous speed, the ray is formed somewhat like a coil or spring, and this spiral nature continues as the ray travels far distances.

Interrelationship among Photons

Isaac Newton believed that a ray of light was made of very minute particles because he said it explained the razor-sharp edge of a shadow. Scientists later suggested that a ray of light consisted of a single strand of continuous stream of energy corpuscles and named these corpuscles ***photons***. There is no existing explanation for the mechanics of placement of this single strand of photons or how a photon in a ray of light maintains its continuous alignment with the photon before and after it. For instance, if the photons in the ray of light were lined up against one another, like a row of golf balls touching each other on a table, the photons would disperse during the travel of a ray in space or in the event of light reflection.

Experience has shown that the photons remain together and do not disperse. Therefore, these particles in the ray of light must have some linkage, and the connection must have a good range of mobility and stability. Hence, it appears that each photon would have to be a round particle with a ball and socket mechanism (like a shoulder joint) extending from it. When all the photons are arranged in a ray of light in a single-stranded stream by this ball-and-socket mechanism, it would have the benefit of being stable and mobile.

Relationship between Photons and Heatrons

When we feel the warmth of the sun against our skin, which is responsible: the photons or the heatrons? As we discussed earlier in this chapter, the heatron concentration is responsible for changes in temperature. Why then do we feel heat when a ray of light from a ray of light from the sun beats down on us? To understand this point, we must refer again to the contraption of energy, where we introduced the theory of how an atom creates photons from ratons. Heatrons are created during the same process as photons; indeed, they are the first products of the atom. Ratons are first transformed into heatrons, and then the heatrons are enclosed into small packets. These packets are photons. That is why radiant light is accomplished by heat.

Resistance Caused by Ratons

Raton particles are so infinitely minute that if a person were to walk through them, the particles would either pass through him or some displacement with negligible resistance would occur. However, if a ray of light was to traverse through the medium of these particles with its tremendous speed, there would be resistance. A light ray covers a distance of 696.6 million miles per hour, and the photons of a light ray, which travel in a spiral, have a velocity of 100 to 1000 times that speed. Likewise, the photons of a gamma ray would move thousands of times faster than those of a light ray. Hence, it is conceivable that resistance is exerted by the ratonic medium, which, in turn, would result in stretching the spirals of a ray.

Furthermore, this theory explains why the intensity of a light bulb is affected by distance. For instance, the light from a lamp appears brighter to a person when he is standing near it, and as the person goes farther away from it, the intensity diminishes. Similarly, a radiant light ray that begins its journey from trillions of light years away reaches the Earth as nearly a straight line, like a radio wave. Thus, it is likely, that at a point, due to its reduced motion, the strands of heatrons that are clumped together to form photons are no longer stable and will eventually break down into the basic particle raton. Even light eventually gets destroyed and is, therefore, of mortal species.

2

Evidence of the Raton in Past Scientific Theories

In the past, scientists assumed that there was some medium for the propagation of light or electromagnetic waves in space. Christian Huygens, a Dutch astronomer of the seventeenth century, called it ether, also known as "illustrated ether."

James Clerk Maxwell, a famous Scottish physicist who formulated equations for electromagnetic waves that propagated through space at the speed of light, said the following:

Whatever difficulty we may have in forming a consistent idea of the constituents of the ether, there can be no doubt that the interplanetary and interstellar spaces are not empty but are occupied by material substances or body which is certainly the largest and probably the most uniform body of which we have any knowledge.

In 1887, two American scientists, Albert Michelson and Edward Malory, conducted an experiment that disproved the ether theory (*see Discussion-3*). In 1905, Albert Einstein formulated the theory of relativity, which assumed that the speed of light is universally constant. This theory further weakened the ether hypothesis, which since then has become nothing but history.

DISCUSSION-3

An experiment done by Albert Michelson and Edward Malory was based on the idea that if the Earth was moving throughout ether, a light beam traveling upwind would be slightly lower than the light beam traveling crosswind. The instrument they devised consisted of reflecting mirrors, a light source, and a reflected light-receiving telescope. The experiment revealed that the upwind light beam and the crosswind beams reached the destination at the same time. The results of the experiments were interpreted as a denial of the ether theory.

The ratonic hypothesis endorses the ether theory and supports Dr. James Clerk Maxwell's notion of the propagation of electromagnetic waves through ether. I disagree with Albert Michelson and Edward Malory and the results of their experiment, which were influential in nullifying the ether theory. There are some considerations that should be taken into account regarding the Michelson-Malory experiment. The length of the light beam that traveled in the experiment was 22 meters. Light travels at the speed of 186,000 miles per second. Thus, light will travel 22 meters in one 1/43,640,000 of a second. The planet Earth is orbiting the sun at the rate of 66,000 miles an hour. In 1/43,640,000 of a second, this planet will travel 11/5,000 of an inch. Furthermore, due to the hypothesis that the raton's (ether's) mass is diminutive to such an extreme that it can permeate the substance of Earth, it follows that there would be little or no displacement of the ratons (ether) during the Earth's passage through the sea of ratons.

No matter what direction a light ray is headed, the resistance is universal and completely independent of the motion of the planet. In view of the above considerations, the result of the Michelson-Malory experiment is not convincing. Furthermore, because of this resistance, the speed of light gradually reduces as it travels through multiple light years. Thus, Albert Einstein's theory that the speed of light is universal stands revised.

Energy to Matter to Heat

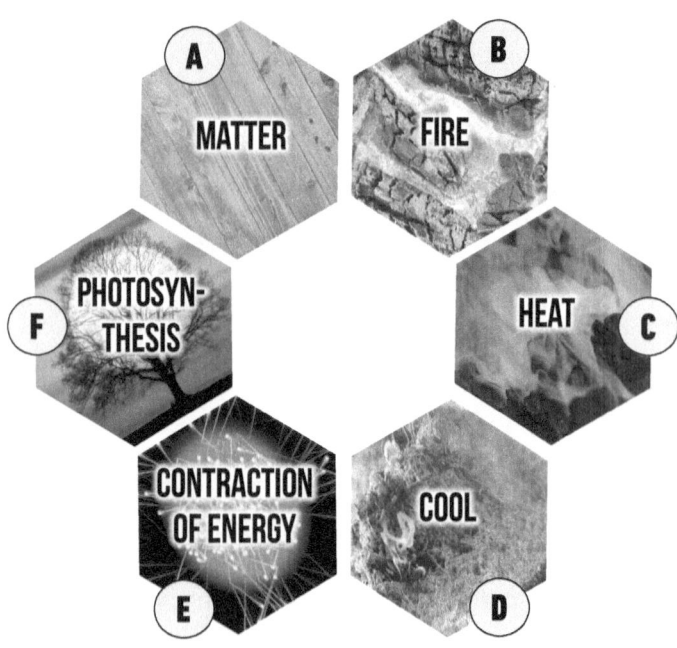

A) Matter – in form of wood
B) Fire – wood burning
C) Heat – fire particle
D) Cool down of heat – Raton particle
E) Contraction of energy in sun – Photon particle
F) Photosynthesis - Back to Wood - Cycle is complete

Part Two

Understanding the Universe

3

The Earth – It's Alive

We have already discussed how matter transforms into ratons and how ratons transform into energy. Now we will explore how energy transforms into matter.

As we mentioned in an earlier chapter, matter on Earth is divided into two categories. One category represents matter derived from photosynthesis. The second category represents matter derived from "geosynthesis," which will be defined as the production of new Earth mass. There is no question that photosynthesis matter is living and dependent on the sun for survival. The Ratonic Theory suggests that matter derived from geosynthesis is also dependent on the sun.

I propose that the Earth is a living entity. To consider this possibility, we must first examine our definition of life. The significant criteria of what constitutes life are: (a) that which is capable of metabolizing, (b) that which is capable of propagating itself. It is currently the consensus that people, animals, plants, and microorganisms are living, but the Earth is not. I propose that the Earth is indeed alive but is not so considered because of the lack of knowledge of its physiology and anatomy.

We have a circulatory system in which the blood, a fluid medium, transports Oxygen from the lungs and nutrients from the gastrointestinal tract to all parts of the body. The Earth has a similar circulatory system, but it is more elaborate. Liquid and gas are channeled throughout the planet in the form of air, rivers, underground streams, ocean currents, rain and much more, which supports the idea the Earth is indeed a living entity.

Earth has a complex metabolic process. To understand this process, it is helpful to use the human body as a comparative model to draw inferences about the Earth's activities such as:

1. Earth mass or geosynthesis
2. Earth deposits, such as petroleum and minerals
3. Earthquakes
4. Volcanoes

GEOSYNTHESIS

Like the human body, Earth utilizes energy for its growth. But unlike our system, which meets its needs by assimilation of food, the terra's source of energy is sunlight. This is like the photosynthesis process by which the plants convert sun energy in the form of sunlight into matter. Earth probably functions on similar principles. It converts the heatron complexes it receives from the photons that arrive on Earth in the form of sunlight. These heatrons make their way from the crust of the Earth to its deepest core. For the production of earth mass, or geosynthesis, to take place, the sun energy (heatrons) is first changed to a primordial earth matter unit, "**earthoblast**." Next, the earthoblasts are differentiated into more mature components by the anabolic process as it utilizes more energy. The final product contains what we know as elements. So far, one hundred and three elements are on record, and the various compositions of these elements make up the planet Earth.

As a result of geosynthesis, Earth's mass and size are increasing constantly. To substantiate this theory, one has to review the early geological events dating back to pre-Cambrian times. Around 500 million years ago, the geographical picture of the planet was entirely different from today's. The Archean rocks of long ago would have appeared enormous relative to the overall size of the planet. Today, however, these rocks are nearly imperceptible within the continental formations. This proves that the planet is far larger today than it has been long ago.

Furthermore, millions of years ago, there existed a huge, sole planet called Pangaea, surrounded by only one ocean. Around 200 million years ago, Pangaea started to split into two separate portions of land, which then further rifted to eventually form the continents as we know them today. These shifts of land mass occurred due to the growing body of the planet.

PETROLEUM

Petroleum, often referred to as black gold, is one of the most important energy sources utilized by present-day civilization. Petroleum is found in deposits in the crust of the planet. How these deposits are formed is not known, although some theories have been put forward, and among them, the organic theory is most accepted (*see Discussion-4*). However, I propose a different theory for the formation of petroleum.

> ## DISCUSSION-4
>
> *The organic theory of petroleum formation states that petroleum is derived from the plankton that covered the prehistoric oceans. The enormous amounts of organic materials that accumulated from the decaying plants and small animals over millions of years were compressed into rock layers. Over millennia, the decaying products underwent organic changes, and the final product is petroleum.*

My theory for the formation of petroleum is based on the premise that the planet is living and functions on principles similar to other living entities, like humans. Humans assimilate food, which is converted to form body mass. If excessive food is consumed, that which is not used by the body is converted into yellowish, spongy tissue and stored mostly under the skin as fat cells. It appears that the planet has a similar converting mechanism. Like humans, the planet has a system of storing unused energy in "fatty tissue." The unused energy converts to a dark black semi-liquid that we know as petroleum. The petroleum is deposited in the outer layer of the crust. The exact mechanism of this secretion is obscure.

As we know, the human body has many secretions, such as gastric juice, saliva, bile, and perspiration. Each of these secretions serves a specific function. In the same way, the planet also has various other secretions besides petroleum, which are not yet defined. These secretions serve as a basis of sending minerals to plants for foliage, fruits, color, fragrance, and more. Earth scientists may explain these secretions as imbibition, but our thoughts should be directed to the in-depth mechanism that directs these secretions.

Our bodies are not designed by scientists, nor is our planet. Scientists have attempted to define life; where it begins, where it ends, and what it is. The criteria for life set by scientists is now a law of nature, but simply a guidepost that is subject to modifications as our knowledge expands. If we can see the similarities between planets and other living entities, we can better understand the phenomenon of life. With our increased understanding, we must expand life's definition to include some things previously thought of as inanimate.

MINERALS

The metal ores and the minerals found deep in the crust of the Earth are the second generation of the substance of the Earth. Like the baking of bricks, the metal ores get baked underground. They are a more-concentrated form of earth substances, and scientists classify them by atomic weights. There is also a third generation of these substances that are radioactive that have still higher atomic weights.

Mines may be compared with trees. Trees are concentrated energy and grow in the outer surface of the Earth. Mines are also high concentrates of energy, except they grown inward. Perhaps one day, geologists may be able to remove portions of a mine that would represent the analogue of a seed, and "sow" it elsewhere. Possibly this seed could become a full-blown mine over a period of thousands of years.

EARTHQUAKES

The spontaneous trembling and shaking of the seemingly stable ground beneath us occurs more commonly than the public tends to realize. An average of twenty tremors a day may be recorded by seismographs in different parts of the globe, the majority of which are so slight that we are not even aware of them. Some are of high magnitude but are confined to regions with no population. Since attention is given only to earthquakes of high magnitude, which occur in highly populated regions, those are the ones that make records in history.

The frequency of earthquakes averages one in every seventy-two minutes, and these events are a result of the Earth's ongoing mass production. Hundreds of millions of years apart, there were great upheavals that brought about anatomical changes in the geographical map. Today, the Earth's circumference is 24,902 miles (4,075 km), which will increase far in the future. Along with the increase in circumference, there will be increased landforms. Currently, we have seven continents, but that number will increase in the far future as more land mass is added.

Earthquakes are seen all over the globe. However, there is an abundance of them around the Pacific Ocean in the area known as the "Pacific Ring." This area encompasses several islands, the most prominent being Japan, the Philippines, Indonesia, and Hawaii. It is possible that the frequency of earthquakes in this area is the result of increased geosynthesis activity occurring deep in the interior of that part of the planet.

The ocean floors have many high mountain ranges, which may also be the result of geosynthesis. One day, there may be a severe earthquake that may cover all the continents. As a result, the ocean floor may stretch, resulting in broader oceans, or the mountains may coalesce to create new land masses. Possibly, the scattered islands of the world would merge to form compact landforms. As new land appears from the ocean, it is also possible that existing land masses could undergo drastic changes.

One can classify the planet as "primary life" and all other living organisms as "secondary life." The secondary life is like a creeping plant that clings to a wall, tree, or ground. Secondary life is dependent on primary life for its survival. When there is severe upheaval and new landforms, all the members of secondary life undergo destruction, nearly to the point of absolute extinction. Then, following millions of years, this secondary life surfaces again with new modes of adaptability to its prevailing events, and a fresh generation of vegetation and animal life marks a new era. Archaeologists believe that this extinction of life must have taken place on our globe at least five times to date.

VOLCANOES

Volcanoes are known to be very vicious, especially when they erupt near populated areas. In 1902, the city of St. Pierre, Nicaragua, with a population of 50,000, was wiped out in a matter of minutes when a volcano erupted in the suburban Mount Pelee. All the inhabitants and buildings were destroyed. These infernos are reputed to be devasting, yet there is a bright side to them.

Volcanoes are crucial for the survival of the planet. Just as in a steam boiler, a pressure release valve releases heat that would otherwise cause an explosion; volcanoes serve as the pressure release valve for the planet. Thousands of asteroids orbit the sun between the orbits of Mars and Jupiter. I propose that in the past, there was a planet in that location that exploded due to accumulated internal heat that failed to be released through volcanoes. These small planetesimals appear to be parts of the planet that exploded, which will be referred to as the "Intermediate Planet" and discussed later.

The expulsion of volcanoes' fiery, hot, molten matter serves two purposes for the planet. First, the volcanoes serve as a means for the planet to excrete its internal waste, similar to the elimination of waste from the human body. It is interesting to note that even certain chemicals that exist in industrial wastes and human excreta exist in volcanic lava such as sulfur compounds. In the process of geosynthesis, the planet functions like a huge industrial plant, and an enormous amount of matter is being manufactured from the Earth's energy. As a result, waste products area produced and must be eliminated through volcanoes.

Second, volcanoes serve the purpose of maintaining the internal temperature balance of the planet, as does the human body. The human body maintains its temperature at 98.6 degrees through various ways, such as perspiration, respiration, urination, and defecation. Similarly, the Earth maintains its temperature through the extrusion of lava.

4

The Genesis of the Solar System

The human body is made up of billions of cells, each with its own ability to metabolize and reproduce. Each cell is an independent living entity, even though it depends on the whole system of the body for its survival. Likewise, living organisms are dependent on the solar system for their survival. From the cell's point of view, it is alive, but the human system is not. Likewise, humans believe that we are living, but the solar system upon which we depend is not living. Planet Earth is living, and its continued life is necessary for the survival of humans and all other life forms.

The other planets of our solar system do not contain life forms like that on Earth. This does not, however, rule out the possibility that they are living. If there is volcanic activity, we can surmise that the metabolic process is functioning.

As discussed earlier in this section, living things are those that (a) utilize a metabolic process and (b) reproduce. We have explored the methods by which our planet utilizes heatrons as food or energy to create its mass and how our planet expels wastes through volcanoes. This process of anabolism and catabolism shows that Earth does indeed metabolize and satisfy the first criteria of life. But what about the second criterion: do planets reproduce? I believe that they may in fact procreate in ways that are beyond our current comprehension.

I hypothesize that planets procreate or propagate and that the sun is the hermaphroditic parent of all the planets in the solar system. Before I elaborate on my views, I would like to review the conventional theories regarding the birth of the solar system. Among the popular scientific theories are the nebular theory, the planetesimal theory, the gaseous (or tidal) theory, the double star theory, and the condensation theory. (*See discussion-5*).

DISCUSSION-5

*According to the **nebular theory**, suggested by French astronomer Pierre Simon Laplace in 1796, the solar system originated from an enormous cloud of hot gas (nebula), which was greater in size than the present solar system. Its center condensed into the sun, and its peripheries, while circling around the sun, sectioned into several rings which ultimately contracted into planets.*

*The **planetesimal theory**, proposed in 1905 by Forest Moulton and Thomas Chamberlain, American astronomer, and geologist, respectively, states that the gravity of a star drew streams of hot gas from the sun as the star quickly passed the sun. As the matter contained in the streams eddied around the sun, some of the particles collided and formed tiny planets, or planetesimals. Some of the planetesimals accumulated large amounts of mass with time, thus forming the planets.*

*In 1919, English scientists Harold Jeffreys and Sir James Jeans argued, in the **gaseous (or tidal) theory**, that the gas that was drawn out of the sun by the gravity of a star, as stated in the planetesimal balls.*

> *The balls of liquid eventually cooled down and developed a solid crust on its perimeter, thereby becoming planets.*
>
> *Then, in the 1930s, the **double star theory** was brought about by an English astronomer named R.A. Lyttletown. He said that the sun had a geminate star, therefore forming a double star. The fellow star combusted into a huge gaseous mass, which, attracted by the sun's gravity, eventually turned into planets, similar to the way presented in the gaseous theory.*
>
> *Finally, the condensation theories evolved during the 1940s to 1950s and are upheld by many scientists. These hypotheses indicate that the few remains of a combusted star, while most of the combusted matter was lost and space, formed a nebula of gas; but in this case, the nebula was drawn in by the sun's gravity. Once again, it is believed that eddies of gas formed into liquid balls, which went on to form solid crusts or their perimeters and become planets.*

My theory suggests that many billions of years ago, the sun started its existence as a moon. For clarification, let us call it Moon-S. Moon-S was beset by another planet, the parent planet. The parent planet attained "starhood," and Moon-S became bigger and evolved into a planet over a period of billions of years. Let us say that Moon-S grew into Planet-S. This Planet-S orbited the parent star for many billions of years and meanwhile propagated several satellites or moons. One of the moons would eventually become the planet Earth. After Planet-S and its moons orbited the parent for billions of years, gradual changes began to take place in their interiors. The thermonuclear furnace started activating, and Planet-S changed to a protostar, which we will call Protostar-S. The protostar continued to orbit the parent star until its generator system in the thermonuclear furnace started to function at full capacity. By then, the Protostar-S became capable of being a full-fledged independent star. At that time, the Protostar-S followed the "routine star ritual" like is parent star and grandparent star. Protostar-S and its offspring left the parent star and traveled to a new location in the casinos.

By this time, Protostar-S attained starhood and became our home star, the sun. The eleven offspring became the planets of our solar system, one of which was the Intermediate Planet that exploded. Another is our moon which failed to thrive, probably due to genetic aberration, and took the role of the Earth's satellite.

The moons that orbit Jupiter, Saturn, Uranus, and Neptune are probably the offspring of the planets that they orbit. Jupiter, Saturn, Uranus, and Neptune are probably on the way to attaining starhood.

Let us take Jupiter as a model and consider its role in the far future. Jupiter has sixteen moons orbiting around it, each of the moons being its offspring. It is a huge planet; its mass is three hundred times that of Earth, and its diameter is ten times that of Earth. Jupiter is following in the footsteps of Planet-S and will eventually "mature" and will be ready for the next phase of its life cycle. The thermonuclear furnace will begin to function or is already functioning; then, its performance will be upgraded to that of a protostar. Due to the increased activity of the thermonuclear generator, Jupiter will generate trillions of miles, and it will continue orbiting the sun for millennia in a binary relationship. In this state, a star matures and will eventually fly out of its orbit.

Let us consider what effects Earth would experience when Jupiter shines like the sun. First of all, there will be no night when Earth is positioned between Jupiter and the sun. Second, the heat will increase significantly, and it may not be possible for Earth's living entities to survive. Third, Earth may start to grow at a faster rate and accelerate its own evolution.

When Jupiter grows to become a full-blown star, it will leave its parent star, the sun, and its solar family to travel to another location in the cosmos, probably somewhere in the spiral galaxy. It will take all its offspring along to its new abode and start a solar family of its own and Jupiter will be the "sun" of that new solar system.

This is the way all the stars in the heavens started their lives. There are more planets than stars in the cosmos, and the life of the universe is infinite, as Fred Hoyle suggested in the Steady-State Theory (see Discussion-6). There are many galaxies in the cosmos, and it is not unreasonable to believe that other planets in the cosmos support life like that on Earth. Some of those civilizations may even have better technology and perhaps are as interested in contacting us as we are in contacting them.

DISCUSSION-6

Big Bang Theory: In the 1940s, Russian-born American physicist George Gamow, along with his associates, formulated the modern version of the Big Bang model for the evolution of the universe. According to this notion, about 10 to 20 billion years ago, there was a spontaneous cataclysmic explosion (a big bang) of

> *the primordial nucleus. The matter thrown out of the exploded nucleus was unimaginably hot and was the harbinger of the universe. It rapidly expanded and started cooling off. In the first one-millionth of one second, elementary particles, such as electrons and positrons appeared. As the matter further cooled, elements such as helium and hydrogen appeared. In the next billion years, galaxies showed up, and following 6 billion more years, our solar system was born.*
>
> ***Steady State Theory:*** *Later in the same decade, English physicist Fred Hoyle and his colleagues formulated the Steady State theory in response to some questions regarding the Big Bang theory, suggesting that the universe has no definite beginning and no end in time. Accordingly, the matter is continuously being created, and new galaxies are continuously being formed. After two decades of debate, the general consensus among scientists favors the Big Ban approach.*

The human population on Earth is over five billion and is growing constantly. Likewise, the stars and planets are multiplying within the universe, thereby explaining why the galaxies are becoming more and more crowded. Generations of stars have lived, and the life span of the universe is inconceivable. If you look up towards the night sky, the panoramic expanse is studded with twinkling stars, narrating the secrets of the elusive disguised entity known as life.

5

The Final Events of the Solar System

In nature, there is nothing that we can classify as immortal. The universe is an ever-changing entity, and the time frame of its alterations, when measured by the human calendar, is so extensive that the scientists studying it are lost in their visualizations, calculations, and speculations. Yet, just like every other mortal, planets and stars will inevitably reach their demise one day. It is the genetic makeup of each star or planet that determines its ultimate destiny.

It is likely that the outer gaseous planets, Jupiter, Saturn, Uranus, and Neptune, will leave one by one with their offspring as they attain adulthood and become nova stars. The inner rocky planets, Mercury, Venus, Earth, and Mars, probably have shorter lifespans and may be destroyed when the sun moves through its final stage of life. Pluto and the asteroid belt may share the same fate as the inner rocky planets.

Scientists have estimated the sun's age to be 4.6 billion years and have projected that the sun will exist as we know it for another 5 billion years. After that time, the sun will end in a catastrophic quietus. Before its demise, the star will undergo debilitating sicknesses. It will swell to a great extent and be a red giant. Then, it will evolve into a supernova when its heat-regulating mechanism ceases to function. This will cause the sun to lose its contours, and its contents will scatter. When the sun's thermonuclear furnace ceases to function as the energy generator, it may be left over as a star remnant. This remnant, depending on its characteristics, will be recognized as a white dwarf, pulsar, or neutron star and will eventually end up in a black hole. These star remnants will next enter the final phase of the star's life cycle and end in the quasar state.

RED GIANT

The point at which a star turns into a red giant is the initial visible indication that the star is on its way out. The star starts to swell and continues to do so until, eventually, it reaches a diameter of up to 240 million miles. Its temperature may exceed millions to billions of degrees Kelvin. I believe that when the sun becomes a red giant, the rocky planets will become engulfed by it and will then reach their respective OTs and convert to flames or firons.

SUPERNOVA

Once a star's heat production mechanism ceases, no further star configuration is retained. The fire, gases, and other remnants of the red giant are spread in an irregular pattern as though a fluid is spilled. The firons start breaking down to Oxygen and heatrons. The heatrons gradually cool down and return to the resting phase, that of the raton. The gases continue drifting in the universal space, leaving the core of the star behind.

The core is the part of a star that functions as the thermonuclear apparatus, which is responsible for the energy production and gravitational force of the star. It is made up of tempered alloy that can withstand infinite heat without undergoing any change. It is radioactive and has a high atomic number.

Different stars have different sizes and internal structures. Some will have cores that become white dwarves, which have some luminosity. Some will become neutron stars, which have a circumference of about twelve miles and high density. Others will be pulsars, which spin at great speeds. All will eventually end up becoming black holes.

BLACK HOLE

After the star's death, it retains its gravitational properly. The gravitational force remains nearly the same, which is high in consideration of its decreased size. This small mass of intense gravitation is the *black hole*. Per cubic millimeter, the black hole will be a billion times denser than that of the original mass of the star.

The gravitational pull of the black hole is the subject of a great deal of scientific literature. Some of the literature suggests that the black hole is the center of the galaxy and that the supergiant is being swung around it. This reminds me of a myth revolving around the octopus. It is said that these strange creatures had the power to wrap their tentacles around ships and disable them. It is believed that if a light beam were to hit a black hole, it would not be reflected back because of its immense gravitational attraction. According to the Ratonic Theory, when the light beam comes in contact with the surface of the black hole, it shatters apart into its component parts, and thus, the light is changed into raton particles. Thus, the light no longer exists, and there is nothing to be reflected back.

Black holes are not permanent. After a star is destroyed, the black hole is left in open space without the protective covering of the million degrees Kelvin temperature it was once accustomed to. Also, they are highly radioactive and lose their radiation with the passage of time. Hence, at some point, external influences or internal factors cause the black hole to undergo spontaneous autolytic combustion, exude a sudden burst of light and transform into an extremely bright object called a *quasar*.

QUASARS

Astronomers once traced the thrusts of radio waves and searched for the possible source of these waves. They came across an extremely bright starlike object and called it a quasi-stellar radio source, or quasar. The first quasar was called 3C-273, and since then, thirteen hundred quasars have been recorded.

Quasars generate an enormous amount of energy in an extremely small area. One explanation for this enigma is the central black-hole hypothesis. According to this theory, the black hole is the center of the quasar. Gas from the surrounding stars in the galaxy is attracted by the intensified gravitational force of the black hole. This gas begins to encircle the black hole, becomes extremely hot, and thus emits X-rays, ultraviolet waves, and radio waves. In this way, the source of energy of the quasar comes from the matter attracted to the black hole.

According to the ratonic hypothesis, the quasar is basically an enkindled black hole. Due to the combustion as discussed earlier, the very makeup of the black hole serves as the fuel for the production of energy of the quasar.

Astronomers who have been studying the intensity of the radio waves of quasars have consistently seen a shift to red, indicating that the quasar's intensity is weakening. This shift has led scientists to conclude that quasars are moving outwards at the speed of light. Because quasars are constantly spreading outward, it has been suggested that the universe is expanding in all directions at the speed of light and that this has been going on for billions of years.

The ratonic hypothesis defies this notion and suggests that quasars are neither moving outward nor is the universe expanding at the speed of light. Instead, the Doppler shift is attributed to a decrease in brightness that is part of the natural annihilation process of quasars. With the passage of time, a quasar loses a great deal of its radioactive energy and mass through the emission of its radiant rays. It gradually contracts until finally it diminishes, much like a smoldering piece of coal.

The fast-moving radiant rays travel long distances through the resistance in the ratonic ocean. Hence, the rays lose their spirals and turn into radio waves by the time they reach Earth. Ultimately, these radio waves will break down and transform back to their resting phase and become ratons.

6

The Asteroid Belt

The word *asteroid* is derived from the Greek words *aster*, which means "star" and *eiods*, which means "like," and coined by Sir William Herschel, a British astronomer. There are innumerable asteroids spread around two million miles orbiting the sun. These asteroids are collectively called the asteroid belt and are located in between the orbits of Mars and Jupiter. On January 1, 1801, Giuseppe Piazzi, an Italian astronomer, was the first person to spot and identify an asteroid. The asteroid was later named Ceres. Ceres was about a thousand kilometers in its largest dimension and is the biggest of all the asteroids in the belt. Earlier, Titus and Bode had formulated the Titus-Bode law that predicted that there should be a planet between the orbits of Mars and Jupiter. The finding of Ceres supported the Titus-Bode law.

As time passed, more asteroids were seen in that region of the solar family, and now there are nearly four thousand asteroids on record. They are of various sizes. Their orbital periods vary with each other, with the vast majority completing a revolution between three to six years. The rotational period around each asteroid's own axis ranges from two to three hours to forty-eight days, but the majority of them would be around four to twenty hours. They are of various spherical, elongated, and kidneylike shapes; their light amplitude is around 0.65. It is estimated that the total combined mass of all the asteroids is about ten percent of the moon's mass.

Theorists once asserted that there must have been a planet in place of the asteroids and that it exploded, but the present hypothesis indicates that Jupiter's enormous mass had influence over the minor planets. The current theory states that Jupiter hindered the planetesimals from merging with each other and drew a large amount of the loose matter that would have been needed for the formation of a planet. It also kept the remaining asteroids from joining together.

The ratonic hypothesis offers strong support for the old belief that there was once a planet orbiting between Mars and Jupiter. I oppose the notion that any planet is formed through the coalescing of planetesimals, which are the derivatives of exploding nearby stars. My objection is based on the belief that the planet is a living wonder. Just as we cannot collect different parts of a human body and stitch them together to create a new functioning living body, it is not possible for fragments of stars and planets to coalesce together to function as a new planet. In nature, every living entity is procreated by another homologous living entity, even though the process of procreation may differ from one entity to another.

The possible planet that existed many millions of years ago was the tenth planet in the solar family. It orbited in the area now known as the asteroid belt. This planet is the Intermediate Planet as mentioned earlier, which failed to survive its youth due to some genetic aberration that kept it from reaching adulthood.

There are various ways in which a planet can come to an end. In the case of the Intermediate Planet, it was probably due to the cataclysmic blast judging from the scattered planetesimals in the asteroid belt.

The possible mechanics of such an ending of the juvenile planet could have been due to the progressive increase of the intraplanetary thermal pressure with no system of pressure release, such as a volcano. As this internal pressure must have accumulated for a long time, it eventually must have reached a point at which it surpassed the capability of the planet's containment, and that could have caused the blast.

BLAST AND THE AFTERMATH

We are familiar with nuclear bombs that can generate enormous heat. One hydrogen or neutron bomb can exert a pressure of over 10,000 tons of dynamite. An exploding planet is able to release billions of times more energy. The basic principle of radiant energy is the compression of heatron complexes and sudden release. Both the nuclear bomb and the exploding planet generate radiant heat on the same principle.

There must have been an enormous shower of outward-shooting meteorites when the Intermediate planet exploded. All of the planets in the solar family must have been impacted by the meteorite shower. Besides the solid leftover asteroids, there must have been a great deal of dust hovering in space for millions of years. Currently, that dust is no longer seen around the asteroids. Since the universe is not static, the dust particles were probably shifted in a current of the ratonic ocean. Scientists have seen great accumulations of dust in various parts of the cosmos. The Milky Way is the most well-known example of such an accumulation. These dust accumulations are probably the result of numerous planets or stars having undergone a demise similar to that of the Intermediate Planet.

Bibliography

Blatt, Frank J. *Principles of Physics*, second ed. Boston: Allyn & Bacon, Inc., 1986.

Constable, George, ed. *Voyage through the Universe: The Cosmos*. Virginia: Time-Life Books, 1988.

———. *Voyage through the Universe: The Stars*. Virginia: Time-Life Books, 1988.

———. *Voyage through the Universe: The Sun*. Virginia: Time-Life Books, 1990.

Flaum, Eric. *The Planets: A Journey into Space*. New York: Crescent Books, 1988

Haber, Heinz, *The Walt Disney Story of Our Friend, the Atom*. New York: Simon and Schuster, 1956.

Hawking, Stephen H. *A Brief History of Time: From the Big Bang to Black Holes*. New York: Bantam Books, 1988.

Raup, David M. *The Nemesis Affair*. New York: W.W. Norton and Company, 1987.

The Human Body – "The Eye: Window to the World." New York: Torstar Books, 1984

The New Encyclopedia Britannica. Marcopaedia, volume 27. Chicago: Encyclopaedia Britannica, Incl., 1989

The World Book Encyclopedia, volumes 2, 6 & 15. Chicago: Field Enterprises Educational Corporation, 1974.

The World Book Encyclopedia of Sciences: The Heavens," volume 1. Chicago World Book, Inc., 1984.

About the Author

DR. BASHAMBER N. CHABRA, has lived a remarkable life shaped by resilience and a thirst for knowledge. Born in Myanmar, he lived through the tumultuous times of World War II, the partition of India and a military coup in his home country, rebuilding his life each time adversity struck. Driven by a desire for a better life, he ventured to Malaysia and later to the USA with his family, embracing new beginnings as a physician.

Throughout his medical career, Dr. Chabra never lost his fascination with the mysteries of the universe. An avid reader of physics and astronomy, he often delved into research during his spare moments. His daughters fondly remember catching him barefoot on the sun-filled patio of their Los Angeles home, contemplating the heat absorbed by the ground. "Dad, what are you doing?" they would ask. While pointing to the warm earth and with a twinkle in his eye, he would respond, "Where does this heat go after the sun goes down?" This curiosity about the fundamental workings of the world fueled his lifelong quest for understanding.

Today, at 93, Dr. Chabra resides in Los Angeles with his wife of 64 years, surrounded by their loving family. His dedication to knowledge and sharing ideas has led him to re-release his book, originally published in 1998, aiming to reach a wider audience in the digital age. Through his work, he invites readers to explore and expand upon the concepts he has passionately researched and pondered for decades.

www.ingramcontent.com/pod-product-compliance
Lightning Source LLC
Chambersburg PA
CBHW030446220526
45464CB00006B/2431